Optimal Reliability-Based Design of Structures Against Several Natural Hazards

Optimal Reliability-Based Design of Structures Against Several Natural Hazards

Alfredo H.-S. Ang, David de Leon Escobedo, and Wenliang Fan

CRC Press
Taylor & Francis Group
Boca Raton London New York

CRC Press is an imprint of the
Taylor & Francis Group, an **informa** business

CRC Press/Balkema is an imprint of the Taylor & Francis Group, an informa business

© 2022 Taylor & Francis Group, London, UK

Typeset by Deanta Global Publishing Services, Chennai, India

Library of Congress Cataloging-in-Publication Data
Names: Ang, Alfredo Hua-Sing, 1930- author. | De Leon, David (De Leon Escobido), author. | Fan, Wenliang, author.
Title: Optimal reliability-based design of structures against several natural hazards / Alfredo H-S Ang, David De Leon, and Wenliang Fan.
Description: Boca Raton : CRC Press, [2022] | Includes bibliographical references and index.
Identifiers: LCCN 2021020707 (print) | LCCN 2021020708 (ebook)
Subjects: LCSH: Structural design. | Structural failures--Prevention. | Reliability (Engineering)
Classification: LCC TA658 .A56 2022 (print) | LCC TA658 (ebook) | DDC 624.1/76--dc23
LC record available at https://lccn.loc.gov/2021020707
LC ebook record available at https://lccn.loc.gov/2021020708

Published by: CRC Press/Balkema
                  Schipholweg 107C, 2316 XC Leiden, The Netherlands
                  e-mail: Pub.NL@taylorandfrancis.com
                  www.crcpress.com – www.taylorandfrancis.com

ISBN: 978-1-032-01130-1 (Hbk)
ISBN: 978-1-032-01136-3 (Pbk)
ISBN: 978-1-003-17728-9 (eBook)

DOI: 10.1201/9781003177289

Typeset in Times
by Deanta Global Publishing Services, Chennai, India

Contents

6 Analysis of Results   71

7 Conclusions and Recommendations   73

Preface

The optimal design of structures is a topic with many challenges and opportunities, especially when the demand is caused by natural hazards. The reliability of structural components has been well established based on calibration. However, the reliability-based design of complex structural systems is still under development and there are neither general standards for the definition of overall structural safety nor a global index that represents the system performance, particularly when the system is challenged by diverse natural hazards like earthquakes, wind, and ocean waves. The identification of critical limit states and the likelihood of their occurrence is a key issue when generating recommendations for optimal design. All these aspects, among others, are the focus of several contributions. In particular, the probability density evolution method (PDEM), which is applied in this work, considers the evolution of the probability density function (PDF) of the critical limit state of the system. The fast simulation and other advantages of the PDEM are remarked upon and illustrated in the examples. The epistemic uncertainty on the demand parameter, and its impact on structural reliability, may be treated as the modeling error of the mean-value of a parameter meaning, therefore, that the reliability index becomes a random variable. The optimal design, which corresponds to the minimum expected life-cycle cost, may be identified and some percentiles or confidence levels can be determined according to the conservative degree desired by the owner or investor. The widening of the options for decision making is especially convenient for large structural systems, as opposed to the traditional practice that resorts only to mean values. The failure consequences are explicitly considered and the expected life-cycle cost is also a random variable. Applications are shown for large infrastructure systems under different natural hazards, and for several high-rise buildings in Mexico City under seismic hazard.

Acknowledgments

The extensive and complicated PDEM calculations for all the design cases of the 3D FEM model of the first 15-story building were performed at the Chongqing University in China with the assistance of Dr. Runyu Liu under the direction of Prof. Wenliang Fan. All the information on costs, including the associated estimations of the damage costs and expected life-cycle costs for all the cases of the first 15-story building and the guide for the last three buildings were provided by Prof. David De Leon of the Autonomous University of Mexico State in Toluca, Mexico.

Special thanks to the Master's students Ricardo Alvarez, Marco Ramírez, and Julio Nava, from the Autonomous University of Mexico State, Mexico, for performing the calculations for the last three buildings.

Introduction

1

1.1 BIBLIOGRAPHICAL REVIEW

Thus far, standards for the reliability-based design of structural components such as beams and columns are well-known and were developed on the basis of calibration (Ellingwood & Galambos, 1982). However, each complex structural system is unique, and the lessons learnt do not have sufficient general insight and experiences to generate standards or recommendations. Furthermore, the potential failure modes and the relative significance of the consequences of structural failure, as a system, acquire a special relevance. There are, therefore, opportunities to study these systems in some detail and to develop the basis for guidelines to systematically design complex structures.

Several approaches have been proposed to develop a basis for understanding the reliability of complex systems and the numerical procedures to produce large simulations and assessments in a more efficient way. The topic of reliability-based optimal design has had several surges and, in the most recent ones, the recommendation of best practices to avoid progressive (cascading) failures has been proposed (for example, Ellingwood & Dusenberry, 2005; Ellingwood et al., 2007; Dueñas-Osorio & Vemuru, 2009). Kriging metamodels are used to optimize reliability-based design (Hyeon Ju & Chai Lee, 2008).

Some other works have used the expected life-cycle cost to do design and maintenance optimization of large infrastructure systems (Frangopol & Liu, 2007; Santander & Sanchez-Silva, 2008). The effect of reducing the reliability of a structural system due to the uncertainty of the design parameters has been analyzed (Der Kiureghian, 2008). Similar results are obtained in the case studies presented in this chapter in terms of the epistemic uncertainty on the mean values of some design parameters, especially on the hazard model.

DOI: 10.1201/9781003177289-1

The concept of the principle of preservation of probability and the generalized density evolution equation has been presented (Li & Chen, 2008). Later on, these principles were compiled in a book, with a comprehensive description of the mathematical basis of the density function evolution (Li & Chen, 2009). Additionally, the advances of the probability density evolution method (PPDM) for nonlinear stochastic systems were introduced as some examples of the high potential of the method on the analysis of complex uncertain systems (Li et al., 2012).

Resilience and sustainability have emerged as new concepts, and are driving risk-based design (Renschler et al., 2010; Francis & Bekera, 2014; Dong & Frangopol, 2015; Lange & Honfi, 2017). Code-calibration practices in oil industry facilities have served as proposed design recommendations (Bai & Jin, 2016). A similar exercise is applied here for the case study of a marine platform.

Some studies resort to indicators to measure the performance of the structural systems (Ghosn et al., 2016) and the robustness and resilience of structural systems (Stochino et al., 2019). Other authors are proposing the use of the directional bat algorithm to perform reliability-based design optimization (Chakri et al., 2017). Other applications of PDEM have been developed for the reliability of lifeline networks (Li, 2018) and to quantify the simultaneous aleatory and epistemic uncertainty of basic parameters of structures (Chen & Wan, 2019).

Recently, high-order analytical moments were implemented to achieve the optimization of structural systems (Rajan et al., 2020), and multi-level, multivariate, non-stationary, and random field modeling are modern tools being used to perform fragility analysis (Xu & Gardoni, 2020). As an additional example of its multiple applications, the probability density evolution method was recently used to perform structural optimization considering dynamic reliability constraints (Chen et al., 2020).

As every system is unique, and its performance depends on many variables, there cannot be a uniform standard for its design. A systematic procedure is still under development for the optimal design of a complete system. What is proposed here is a practical probabilistic procedure for the reliability-based optimal design of a structure as a whole system. In other words, the proposed procedure will determine the reliability-based safety index for the optimal design of a complete structure as a system (Ang et al., 2019).

Aleatory and Epistemic

2

2.1 EPISTEMIC UNCERTAINTY FOR OPTIMAL STRUCTURAL DESIGN

In engineering, uncertainties cannot be avoided; however, when engineering systems are exposed to uncertain hazards, adequate modeling of these uncertainties becomes a crucial step on the way towards the optimal design (Ang et al., 2019). In fact, the main objective of the reliability approach is to handle uncertainties in a proper and rational manner. Standards for the structural design of components were developed by considering reliability and calibration exercises, and they constitute an appropriate guide to building and consolidating the basis for the optimal reliability-based design of complex systems (Ellingwood & Galambos, 1982).

Usually, engineering uncertainties are classified into two broad types: the *aleatory* and the *epistemic* types. The aleatory type is recognized as the intrinsic variability or randomness in nature, and may be appraised and quantified through statistical observations, whereas the epistemic type is the imperfect knowledge involved in one's attempt to predict reality. Epistemic uncertainty is usually referred to as the modeling error or the incomplete information to characterize a variable, and will often require subjective judgments to complement the variable representation (Ang et al., 2019).

Due to their different natures, these two types of uncertainty are known, respectively, as "data-based" and "knowledge-based" uncertainties. Being an inherent part of nature, the aleatory type cannot be reduced; by contrast, the epistemic type may be reduced, up to a certain point, with further knowledge of the true state of nature. In practice, it is possible to reduce epistemic uncertainty through additional investments or research, and cost/benefit studies may put in perspective the relative cost of the studies against the gains on reliability and lower expected costs in the design of a large and complex engineering system (Ang et al., 2019).

Aleatory uncertainty is measured through statistics (frequency distributions, histograms, probability density functions, etc.) or in an approximate way, through the moments (mean, variance, third moment, and higher moments) which are obtained in terms of observed data. Epistemic uncertainty is assessed on the basis of the degree of knowledge of the design parameters, mainly on the unknown demands, and subjective judgment is usually necessary. As the mean state of nature is most important, the mean state may be designated within a range of possibilities – representing the epistemic uncertainty. This range can then be translated into an equivalent coefficient of variation (c.o.v) and with a prescribed probability density function (PDF) such as the lognormal (Ang et al., 2019).

Optimal Design of Structural Systems

3

3.1 EXPECTED LIFE-CYCLE COST FOR THE OPTIMAL DESIGN OF STRUCTURES

For the optimal design of onshore or offshore structures, the expected life-cycle cost, $E(LCC)$, is a convenient cost measure (Ang et al. 2019). This should include all the cost of failure or damage consequences over the service life, which is usually 50 years. Commonly, the $E(LCC)$ is expressed as:

$$E(LCC) = C_I + E(C_F) \tag{3.1}$$

where:

C_I = the initial cost, including the design and construction costs, and
$E(C_F)$ = the present value of the expected costs of failure consequences.

Depending on the structure use, the cost components of the failure consequences may include several items, such as:

- Fatalities and injuries, C_{fi}
- Loss of contents, C_c
- Loss of profit or business interruption loss, C_p
- Cost of repairs, maintenance or structure replacement, C_r.

Therefore, the expected cost of failure consequences are defined, for example:

$$E(C_F) = PVF[E(C_{fi}) + E(C_c) + E(C_P) + E(C_r)] \qquad (3.2)$$

where:
PVF = the present value function, expressed as:

$$PVF = [1 - \exp(-rT)] / r \qquad (3.3)$$

where:
r = net annual discount rate, and
T = structure lifetime

A conceptual graph that shows the typical trends of the initial cost, the present value of the expected failure cost, and the expected life-cycle cost is shown in Figure 3.1 (Ang et al., 2019).

In to Figure 3.1, the horizontal axis represents the alternative designs, which have a reliability index β, whereas the vertical axis represents the costs. The initial cost, C_I, grows with the reliability index, because a conservative design requires a more robust and expensive structure. On the contrary, the present value of the expected failure cost, $E(C_F)$, reduces as the reliability index grows. This is because, as the design is more conservative, the failure probability and the expected failure costs reduce and trend to zero. The expected life-cycle cost, E(LCC), decreases up to a certain point where it begins to increase again. Therefore, there is a reliability index that will correspond to

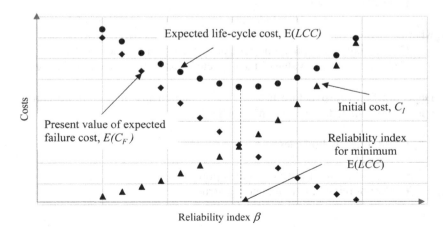

FIGURE 3.1   Costs versus reliability index for respective cost items.

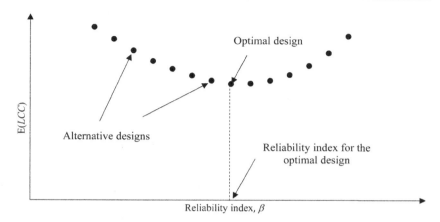

FIGURE 3.2 Expected life-cycle costs E(*LCC*) for several reliability indices β and optimal design expected life-cycle costs E(*LCC*) (Ang et al., 2019).

the minimum cost and will allow the identification of the optimal design (Ang et al., 2019).

To determine the structural design with the minimum expected life-cycle cost, proceed to design the structure with varying design safety indices, β, and estimate the corresponding E(*LCC*). Plot the resulting designs as shown in Figure 3.2, showing the various specific designs. From Figure 3.2, the optimal design with the minimum E(*LCC*) and corresponding β can be identified (Ang et al., 2019).

Calculation of Reliability Applying PDEM

4

A powerful procedure to calculate the reliability of a complex system, the *probability density evolution method* (PDEM) published by Li and Chen (2009) has shown its effectiveness in obtaining the probability density function (PDF) of the performance function Z_{max}, i.e., $f_{Z_{max}}(z)$, from which the mean value of the system reliability, R, can be calculated through the integration over z,

$$R = \int_{\Omega} f_{Z_{max}}(z)\,dz \tag{4.1}$$

where Ω is the safe domain of the system (Ang et al., 2019).

A reasonable way to include epistemic uncertainty is by considering it as the error in the estimation of the mean-value of Z_{max}. A set of all possible values of the mean of Z_{max} will be generated according to the epistemic uncertainty (Ang et al., 2019).

Therefore, the mean-value of Z_{max}, $\mu_{Z_{max}}$, will become a random variable and its PDF may be conveniently assumed to be represented by a lognormal distribution with a mean of 1.0 and a specified coefficient of variation (c.o.v.) (covering an equivalent range of values that corresponds to the dispersion of mean values). Therefore, with the PDF of Z_{max} conditional to the mean value, $f_{(z_{max}|\mu)}(z)$, and with the PDF of μ_Z, or $f_{\mu_Z}(\mu)$, the convolution integration of these two PDFs would then yield the final system reliability (Ang et al., 2019).

$$R = \int_0^\infty \int f_{(z_{max}|\mu)}(z)\,dz f_{\mu_z}(\mu)\,d\mu \tag{4.2}$$

Equation 4.2 may be assessed with a simple Monte Carlo simulation (MCS); therefore, the histogram of the system reliability, or its reliability index, may be built. The random features of the variables, including both types of

DOI: 10.1201/9781003177289-4

uncertainty, may be included in the simulation process to calculate the system reliability. Percentiles of this histogram, for the calculated safety indices, may be related to respective levels of statistical "confidence", i.e. confidence intervals. According to this, any desired confidence level of the safety index may be calculated and high levels may be specified for the conservative design of a complete structural system (Ang et al., 2019).

In fact, when a high confidence level is selected for design, the effects of the epistemic uncertainty may be minimized and additional opportunities to provide safety to a design may be offered, as opposed to the conventional practice of resorting only to the mean value. As described previously, the PDEM has many advantages over the Monte Carlo simulation when assessing the reliability of a complex system (Ang et al., 2019). Although some of these benefits are presented here, as an intuitive description, the method has a complete theoretical support (Li & Chen, 2009).

From an operational point of view, the PDEM is similar to the MCS method because both methods require deterministic samples from the structural response analyses to handle the numerical procedure (Ang et al., 2019). The difference resides in the way the random samples are chosen: the MCS makes an arbitrary, selection whereas the PDEM does it on a mathematical basis, driving the process towards an optimal simulation (Chen & Li, 2007). In addition, each sample corresponds to a probability that is mathematically calculated (Chen et al., 2009).

The formal support of the PDEM is presented by Li and Chen (2009). In brief, the technical background is presented in the next paragraphs (Ang et al., 2019). The probability density evolution equation is based, from Li and Chen (2009), on the expressions:

$$\frac{\partial p_X\Theta(x,\theta,t)}{\partial t} + \dot{X}(\Theta,t)\frac{\partial p_X\Theta(x,\theta,t)}{\partial x} = 0 \qquad (4.3)$$

where the variables in Equation 4.3 are: $p_{X\Theta}(X,\Theta,t)$ is the joint PDF of X, $\boldsymbol{\Theta}$; $X(\Theta,t)$ is the structural response, and $\dot{X}(\Theta,t)$ is the velocity of the structural response, respectively. $\boldsymbol{\Theta}$ is a vector of the random variables pertinent to the structure capacity and demand. The initial condition of Equation (4.3) is

$$p_{X\Theta}(X,\Theta,t)\Big|_{t=0} = \delta(x-x_0)p_\Theta(\theta) \qquad (4.4)$$

Equation 4.3 with the initial condition of Equation 4.4 can be solved using the finite-difference method.

The process for solving Equations 4.3 and 4.4 may be summarized in the following steps (Ang et al., 2019):

1. Generate the "representative or sample points" and their respective probabilities in the solution space. Obtain the deterministic solution of the system response for each representative point and also its associated probability.
2. Determine the joint PDF, $p_{X\Theta}(x,\theta,t)$ using the finite-difference method.
3. By numerical integration, obtain the numerical values of the PDF of Step 2.
4. On the basis of the "complete system failure process" (see Chen & Li, 2007) the reliability of a system is defined by the one-dimensional PDF of the ultimate system performance, Z_{max}; integration of this PDF yields the system reliability as given by Equation 4.1.

The computational implementation of the PDEM is similar to that of the MCS – in the sense that it requires the deterministic solution of the system response for each representative point (or sample point) in the solution space. However, the representative points in the PDEM are selected on the basis of a mathematical algorithm (Li & Chen, 2009) and are associated with respective probabilities; determining these associated probabilities for the respective representative points is central to the PDEM and is a significant and non-trivial mathematical problem (Chen et al., 2009). Experience shows that the number of representative points needed to obtain accurate results with the PDEM is around 200–400 points even for very large and complex systems (Ang et al., 2019). In contrast, the representative points, or sample size, in MCS are randomly generated to cover the entire population of possible deterministic response solutions; in this case, the corresponding number of sample points (with the MCS) can be extremely large compared to that of the PDEM. Other significant differences between the PDEM and the MCS are the following (Ang et al., 2019):

- While the MCS obtains the mean reliability of a system, the PDEM obtains the PDF of Z_{max}. The ultimate system performance and the mean system reliability is obtained through Equation 4.1.
- There is no mathematical rigor associated with the MCS; it is a crude or brute-force random sampling technique. By contrast, the PDEM is based on mathematical rigor as presented in Equations 4.3 and 4.4.

Thus far, for a complex engineering system, the practical assessment of its reliability invariably requires Monte Carlo simulation or its improved versions such as variance reduction. However, there is a limit to the effectiveness of any MCS method, particularly for large and complex dynamic and highly nonlinear systems. The recent development of the PDEM provides an effective alternative computational tool for the required reliability assessments that can and should serve to widen the practical implementation of the reliability approach, particularly for complex engineering systems (Ang et al., 2019).

Through the application of the PDEM and based on the complete system failure process proposed by Chen and Li (2007), the reliability, R, of a system is defined as the *system capacity is greater than the applied load*, and can be obtained through the integration of a one-dimensional extreme-value PDF of the ultimate system performance Z_{max}, as presented in Equation 4.1 (Ang et al., 2019).

It is important to emphasize that the above procedure for system reliability, through Equation 4.1, completely circumvents the need to identify the possible failure modes of a system and their respective mutual correlations which are necessary in any traditional methods (e.g., Ang & Ma, 1981) for numerically assessing system reliability. This is not surprising as the implementation of the PDEM may be interpreted, heuristically, as a "weighted" sampling process similar to the MCS in which the non-trivial weight for each sample is the probability associated with each representative point. Recall that through MCS, the identification of the failure modes is unnecessary (Ang et al., 2019).

The role of the PDEM is especially significant in the assessment of the reliability of a highly complex system. For such a system, the analysis of its reliability would traditionally rely on Monte Carlo simulations; however, because of the large number of degrees-of-freedom needed to accurately model a complex dynamic system, and because of the very large sample size required (of the order of 10^6 for very small failure probabilities) in any MCS in order to achieve sufficient accuracy, it could be impractical or too expensive to apply the MCS (Ang et al., 2019).

With the PDEM, the reliability of a system can be defined through the complete system failure process defined by Chen and Li (2007). On this latter basis, the reliability of a system becomes

$$R = Prob\left[Z_{max} > 0\right] \tag{4.5}$$

where Z_{max} is the ultimate system performance under the load effects; Z_{max} is a function of the system parameters and of the load effects (Ang et al., 2019). The corresponding reliability index is:

$$\beta \approx -\varnothing^{-1}\left[R\right] \tag{4.6}$$

where $\varnothing^{-1}$ is the inverse standard normal distribution.

As stated earlier, the calculation of the PDF of Z_{max} provides a rational and practical basis to include the effects of the epistemic uncertainty into reliability-based design of a complex system (Ang et al., 2019).

Application Examples

5

5.1 DESIGN OF OFFSHORE OIL PRODUCTION PLATFORMS

The industry standards (American Petroleum Institute, 1993; Mexican PEMEX, 2000) for the design of offshore oil production platforms require the reliability index for the design of such production platforms to be within $\beta = 3.3-3.5$.

In the study of life-cycle cost (*LCC*) design of offshore platforms, De Leon and Ang (2008) show the histogram of β for the minimum *LCC* design in Figure 5.1 (Ang et al., 2019). Figure 5.2 summarizes the 90% β for different designs with an optimal 90% $\beta = 3.45$. Clearly, in the case of offshore production platforms, a confidence of 90% is consistent with good professional practice, i.e., within $\beta = 3.3-3.5$ (Ang et al., 2019).

5.2 DESIGN OF CABLE-STAYED BRIDGES

In the case of the design of cable-stayed bridges, consider the particular bridge in Jindo, Korea (Han & Ang, 2008). For the minimum life-cycle cost design of the Jindo bridge, the histogram of the reliability index, β, is shown in Figure 5.3, indicating that the mean reliability index E(β) is 2.28, whereas the 90% β is 3.23 (Ang et al., 2019).

The reliability index in the actual design of the bridge (see Figure 5.4) can be inferred to have a slightly higher mean value than the mean reliability

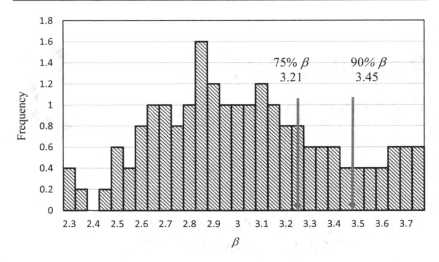

FIGURE 5.1   Histogram *of β* for minimum expected life-cycle cost design.

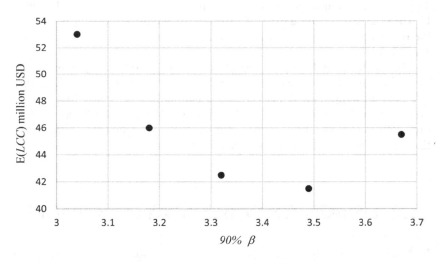

FIGURE 5.2   90% *β* versus expected life-cycle cost E(*LCC*).

index of E(β) = 2.28; thus it follows that the corresponding 90% β would be > 3.23, indicating that the reliability index underlying the actual design could be within 90–95% (Ang et al., 2019).

Therefore, the required confidence level for the minimum expected life-cycle costs E(*LCC*) design of cable-stayed bridges appears to be in the range of

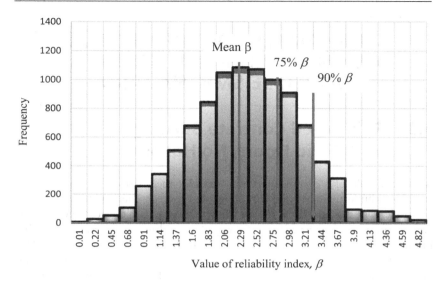

FIGURE 5.3 Histogram of β for minimum *LCC* design of Jindo bridge.

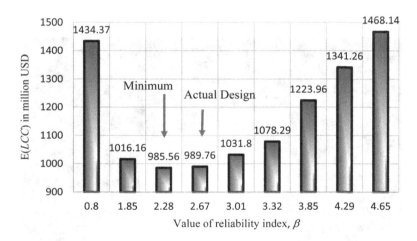

FIGURE 5.4 E(*LCC*) vs β for different designs of the Jindo bridge.

90–95%. In summary, therefore, from the above observations, it appears that for the optimal design of critical engineering systems, a safety index within the 90–95% confidence level is consistent with good professional practice, and may be considered acceptable to ensure safety for the design of a complete structural system (Ang et al., 2019).

5.3 EXAMPLES OF THE OPTIMAL DESIGN OF HIGH-RISE BUILDINGS UNDER SEISMIC HAZARD

5.3.1 Reinforced Concrete Building with Solid Floors, No Hollow

As an example of a complex system, consider the 15-story reinforced concrete (RC) building in Mexico City shown in Figure 5.5; its three-dimensional finite element model (3D FEM) is shown in Figure 5.6. Described below are the properties of the actual 15-story RC building as designed and built. Figures 5.7 and 5.8 show its 2D elevation and plan sections, respectively. The building was subjected to the Mexico earthquake of 1985; the two directions of the earthquake ground motions are shown in Figures 5.9 and 5.10 (Ang et al. 2019).

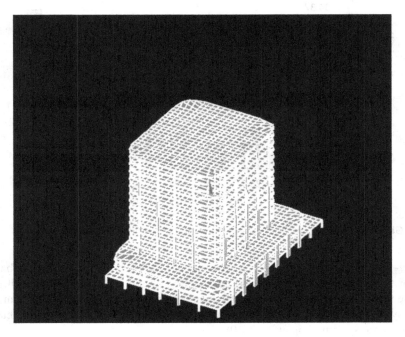

FIGURE 5.5 A 15-story RC building in Mexico City.

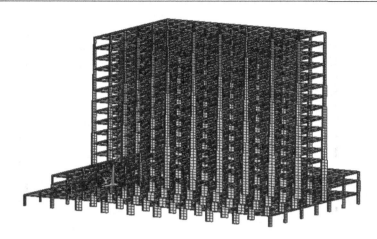

FIGURE 5.6 3D FEM of 15-story building.

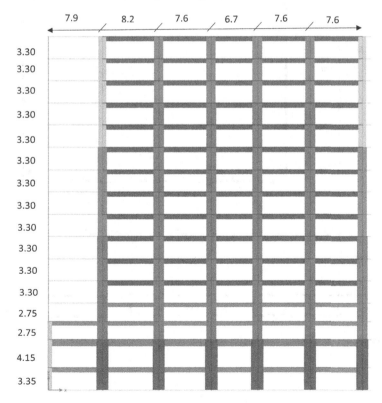

FIGURE 5.7 2D elevation view of the 15-story building (dimensions in m).

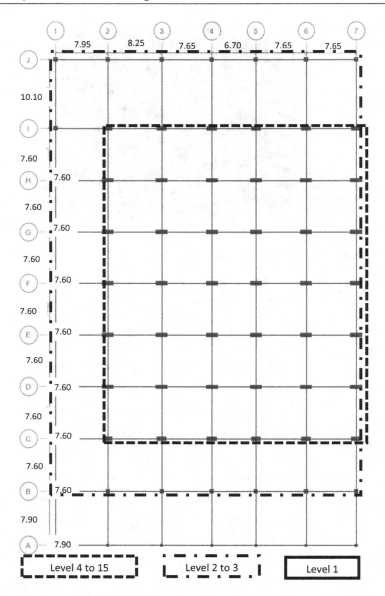

FIGURE 5.8   Plan view (dimensions in m).

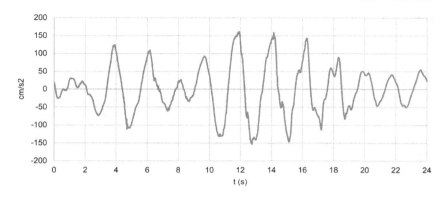

FIGURE 5.9   1985 SCT record, EW Ground Motion.

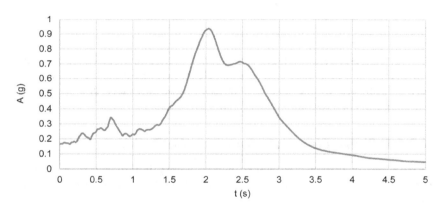

FIGURE 5.10   Design spectrum Mexico City.

The details of the column sizes at the different stories and the corresponding reinforcement re-bars are indicated in Table 5.1 and graphically described in Figure 5.11 (Ang et al., 2019).

The building was designed according to the current seismic regulations in Mexico City (GCDMX, 2014).

In this example, E_1, E_2 are the concrete initial Young's modulus from the basement to level 7, and from level 8 to 15, and and $f'c$ is the concrete resistance, respectively, and are random variables with statistical information listed in Table 5.2. The reliability for the maximum responses can be described by Equation 5.1 (Ang et al., 2019),

$$R = Pr\left\{Z_{max}\left(E_1, E_2, f'c, \tau\right) > 0\right\} \tag{5.1}$$

TABLE 5.1   Details of Column Sizes and Rebars at Various Building Levels

LEVEL	SECTION (cm)	STIRRUP A	STIRRUP B	DETAILS (SEE FIGURE 5.11)	REBARS A	REBARS B
Basement to level 3	150 × 60	# 4 @8"	#3 @10"	1-1	2#10 + 1#8	1#8
Level 4 to level 6	125 × 60	#3 @8"	#3 @10"	1-2	2#10	1#8
Level 7 to level 8	100 × 60	#3 @8"	#3 @10"	1-3	1#10 + 1#8	1#6
Level 9 to level 10	85 × 60	#3 @8"	-	2-1	2#8 + 1#6	-
Level 11 to level 14	60 × 60	#3 @10"	-	2-2	2#8	-
Level 15	45 × 45 and 40 × 40	#3 @10"	-	2-3	1#8 + 1#6	-

in which Z_{max} is the performance function and determined by $X_j(.)$ (j = 1,...,,15), which are the maximum responses for all stories j at time τ over the time period T. Implicit are the capacities for each limit state, according to the current Mexican code (GCDMX, 2014) (Ang et al., 2019).

5.3.1.1 Finite Element Model of the Building

In each of the response analyses of the 15-story building, the building is modeled by 3D finite elements as described, in two perspective views, in Figure 5.12.

5.3.1.2 Response Analysis and Optimal Design of Building

For this building, the calculations of its response for the 3D model of the building is quite involved. The number of finite elements is over 205,000, and the total number of nodes and degrees of freedom (DOFs) are over 140,000 and 800,000, respectively, if the model with shell elements is used. However, the number of finite elements is over 72,000, and the total number of nodes and DOFs are over 63,000 and 200,000, respectively, for the model used in the probability density evolution method (PDEM) analysis without the shell elements.

With the PDEM (Chen & Li, 2007), about 135 representative points or samples of deterministic building responses, with their respective probabilities, were necessary to obtain the probability density function (PDF) of the ultimate

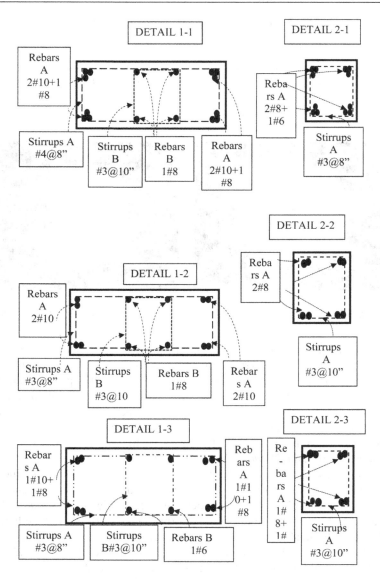

FIGURE 5.11 Graphic details of columns and associated rebars.

TABLE 5.2 Concrete Initial Young's Modulus and concrete resistance ($\times 10^4$ MPa)

VARIABLE	E_1	E_2	$f'c$
Distribution	Normal	Normal	Normal
Mean value	3.25	3.15	0.35
Standard deviation	0.325	0.315	0.09

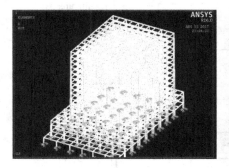 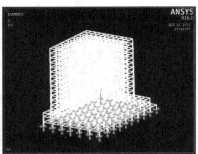

FIGURE 5.12   Two perspective views of 3D FEM model of the 15-story building.

performance function, Z_{max} of the building response as shown in Figure 5.13 (with twin modes); the mean reliability index, conditional to the EW SCT record of the 1985 Mexico City earthquake, is $\beta = 2.63$. This PDF represents only the effect of the underlying aleatory uncertainty. Introducing the distribution fitting procedure, the same aleatory uncertainty can be modeled approximately with the fitted lognormal PDF with a mean value of 0.9 and standard deviation of 0.035 (also shown in Figure 5.11); the corresponding mean reliability index of the fitted lognormal PDF would be $\beta = 2.73$. (Ang et al., 2019).

For this example, assume that the epistemic uncertainty (representing the inaccuracy of the mean-value of Z_{max}) can be modeled also with a lognormal PDF with a mean-value of 1.0 and a coefficient of variation (c.o.v.) of 0.10. With Equation (4.2), convolution integration of this lognormal PDF and the lognormal PDF of the aleatory uncertainty with a mean of 0.9 and standard deviation of 0.035, then yields the histogram of all possible values of the safety index for the 15-story building as shown in Figure 5.14 (Ang et al., 2019).

On Effectiveness of the PDEM – The effectiveness of the PDEM can be demonstrated in this example. In this regard, the following was observed:

- Only 135 sample points, with corresponding associated probabilities, of the building responses were used to calculate the reliability for each case of the building design. For each case, the solution is performed with the Dirac δ sequence method (Fan et al, 2009).

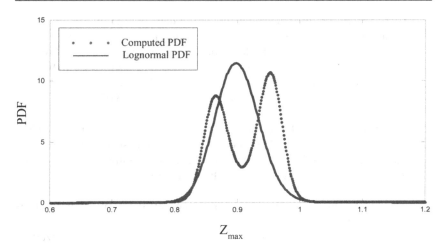

FIGURE 5.13 Computed PDF of Zmax of the building and corresponding fitted lognormal PDF.

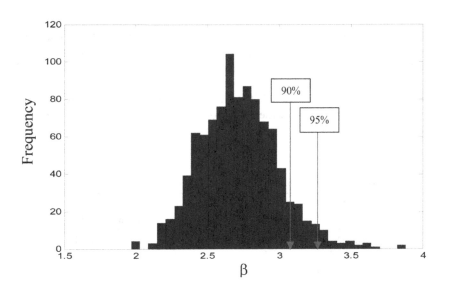

FIGURE 5.14 Histogram of β for the original 15-story building based on lognormal PDF of Zmax.

- To verify the accuracy with this small sample size of 135, calculations with 226 sample points were also performed for the case of the original as-built building design – the results are as follows: with 135 sample points, the range of safety index with 90% to 95% confidence is $\beta = 3.05$–3.18, whereas with 226 sample points, the corresponding range of the reliability index is $\beta = 3.07$–3.20.

The difference between the results for the two sample sizes is around 0.6%, indicating that only 135 sample points for the original case are needed to give reasonably accurate reliability indices for the 15-story building. It is reasonable to assume that the other cases of the same building will have the same accuracy (Ang et al., 2019).

The histogram of the corresponding reliability index of the building is shown in Figure 5.14.

The statistics of the histogram in Figure 5.14 can be summarized as follows:

Mean $\beta = 2.72$
90% confidence $\beta = 3.05$
95% confidence $\beta = 3.18$

Similar PDEM calculations were performed for the same building with different percentages of the original building design; namely, 100%, 105%, 110%, 120%, and 130% of the original as-built structure. The results of all these cases, namely the reliability indices β together with the corresponding expected life-cycle costs E(LCC) in million US dollars (USD), are summarized below in Table 5.3. For the PVF, from Eq. (3.3), $r = 8\%$ and $T = 50$ years (Ang et al., 2019).

TABLE 5.3　Reliability Index β and E(LCC) for All Cases

CASE (% OF ORIGINAL)*	100%	105%	110%	120%	130%
Mean β	2.72	2.85	2.99	3.31	3.74
E(LCC) **	11.24	9.59	8.46	7.34	7.21
90% conf. β	3.05	3.23	3.39	3.71	4.22
E(LCC)**	7.46	6.80	6.59	6.69	7.07
95% conf. β	3.18	3.37	3.55	3.87	4.39
E(LCC)**	6.73	6.36	6.30	6.60	7.05

*in percentage of the original as-built structure
**in million USD

Plots of the reliability index β (for 90% and 95% confidence) versus the corresponding E*(LCC)* are shown below in Figures 5.15 and 5.16, respectively.

Figures 5.15 and 5.16 clearly show that with 90–95% confidence the optimal design of the 15-story building would have required a safety index of β = 3.39–3.55, which is 110% of the original design; i.e., to obtain the minimum E*(LCC)* design would require slightly higher than the original as-built building (110%). Observe from Table 5.3 that the original as-built structure was

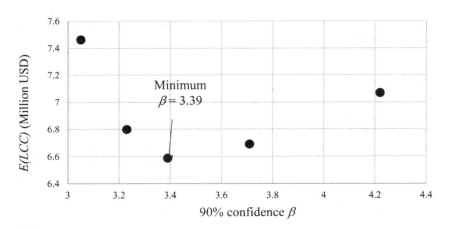

FIGURE 5.15   Plot of alternative designs with 90% confidence β versus respective E*(LCC)*.

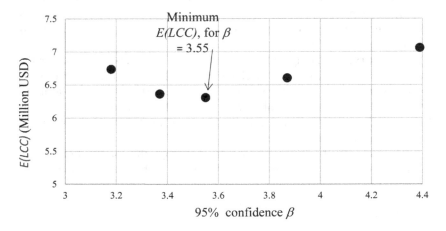

FIGURE 5.16   Plots of alternative designs with 95% confidence β versus respective E*(LCC)*.

designed with a safety index of β = 3.05–3.18 with the same 90–95% confidence (Ang et al., 2019).

5.3.2 Example of Framed RC Building with a Hollow Space

The same building analyzed before is considered now, except that a hollow space almost at the center is added to revise its effects on the optimal design: safety level and costs.

The plan and elevation sections indicate in Figures 5.17 and 5.18, respectively, the distribution of cross-sections for the typical frame indicated in Figure 5.17, is shown in Figure 5.19.

The cross-sections $b \times h$ and concrete strength $f'c$ for columns and beams are shown in Tables 5.4 and 5.5.

Table 5.6 shows the dead loads and Table 5.7 the live load for office buildings, according to current codes for Mexico City (GCDMX, 2014).

The natural periods and modal shapes are calculated; the periods and frequencies for the first 3 modes are shown in Table 5.8, and the first 3 modal shapes in Figures 5.20, 5.21 and 5.22.

A sample of the roof lateral displacements, for the east–west SCT record of the first seven seconds of the 1985 Mexico City earthquake, is shown in Figure 5.23.

A sample of the bending stress ratio, ultimate negative and positive moment ($Mu-$ and $Mu+$) versus capacity (Mr) for beams is shown in Table 5.9.

Interaction diagrams for two critical columns are shown in Figures 5.24 to 5.25 where D/C are the demand versus capacity ratios and the squares represent the maximum demand.

As observed, bending on beams controls the failure, which indicates that the strong column-weak beam mechanism was followed in the design process. A distortion analysis for collapse prevention is performed for each alternative design, and it was found that all of them satisfy the allowable interstory drift, 0.03 according to the current code in Mexico City (GCDMX, 2014). As an example, Table 5.10 shows one of them.

Therefore, the limit state is considered through the combination of local failures at the most critical beams and the failure probability is defined as:

$$p_F = \bigcup_{i=1}^{n} B_i = Pr\left(B_1 \mathrm{U} \, B_2 \mathrm{U} \ldots \mathrm{U} B_n\right) \tag{5.2}$$

where B_i is the failure of the beam i and n is the number of critical beams.

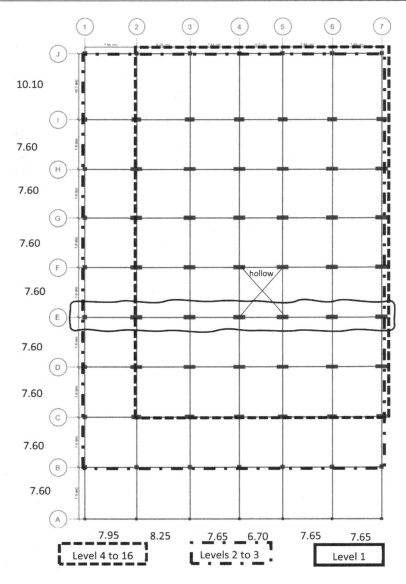

FIGURE 5.17 Plan view, analyzed frame, and hollow space (dimensions in m).

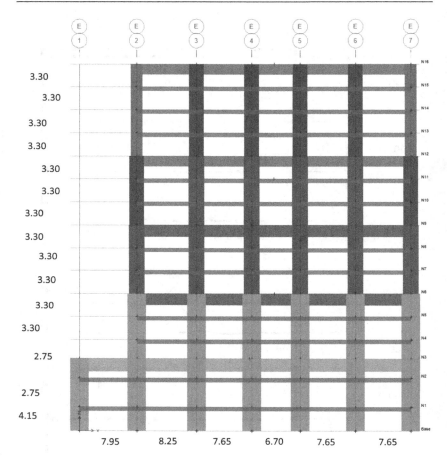

FIGURE 5.18   Elevation view (dimensions in m).

After the calculation of the failure probability, the expected life-cycle costs are appraised. From Eq. (3.3), and the same data as for the last building, $PVF = 12.27$. Five alternative designs were assessed and the optimal design was identified; Table 5.11 shows the results and the optimal design is in bold.

It is observed that the optimal reliability index is 3.51 and the optimal life-cycle cost is 12.79 million USD.

By following the same procedure as for the first building, some levels of epistemic uncertainty are introduced to assess the effect of these levels on the building reliability index and expected life-cycle cost. The histograms for the reliability index and the expected life-cycle cost are shown in Figures 5.26 and 5.27. Expected life-cycle costs for 90% confidence (β) are shown in Figure 5.28.

The mean value, 90% and 95% percentiles are shown in Table 5.12.

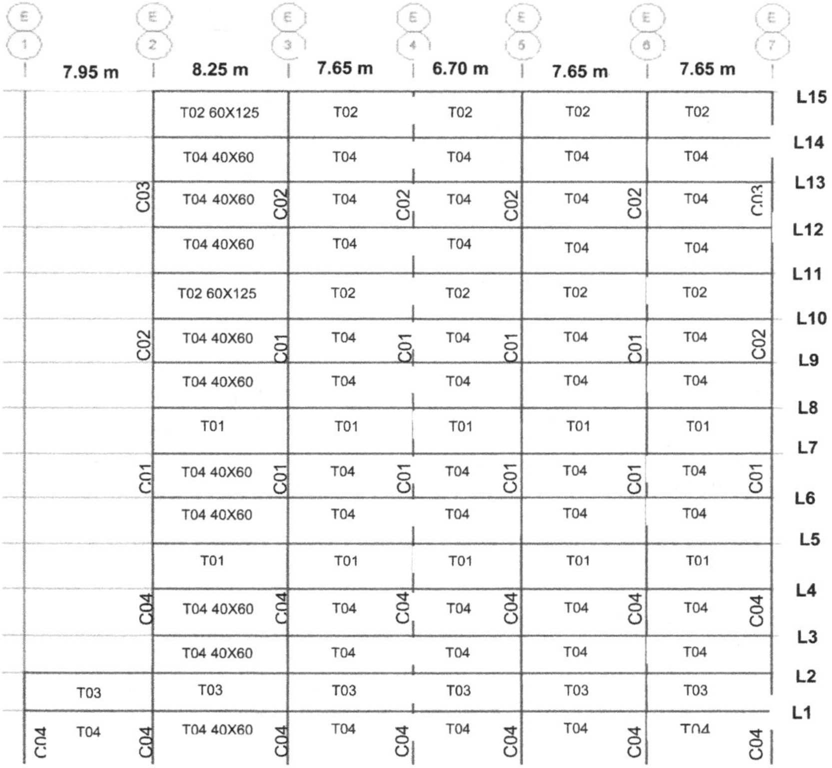

FIGURE 5.19 Members sections for frame 5 (dimensions in cm).

TABLE 5.4 Cross-Sections (cm) and f'c (kg/cm²) for Columns

COLUMN	b	h	f'c
C01	60	175	350
C02	60	150	250
C03	60	135	250
C04	60	115	250

TABLE 5.5   Cross-Sections (cm) and $f'c$ (kg/cm²) for Beams

BEAM	B	H	f'c
T01	40	70	250
T02	45	75	250
T03	50	70	250
T04	55	75	350
T05	55	75	250
T06	60	100	350

TABLE 5.6   Code Dead Load for Interstories and Roof

CONCEPT	LOAD (kg/m²)
Interstories	
Slab 40 cm thickness	430
Floor cover	55
Ceramic	70
Facilities and false ceiling	60
Interior walls	100
Overload (code)	40
Total	755
Roof	
Slab 40 cm thickness	430
Floor cover	55
Waterproofing	10
Bricks	40
Filling	40
Overload (code)	40
Facilities and false cieling	40
Total	655

TABLE 5.7   Live Load for Office Building (Code) (Kg/m²)

MAXIMUM LIVE LOAD	INSTANTANEOUS LIVE LOAD
250	180

TABLE 5.8   Periods and Frequencies for First 3 Modes

MODE	PERIOD SEC	FREQUENCY CYCLE/SEC
1	1.008	0.992
2	0.379	2.64
3	0.204	4.905

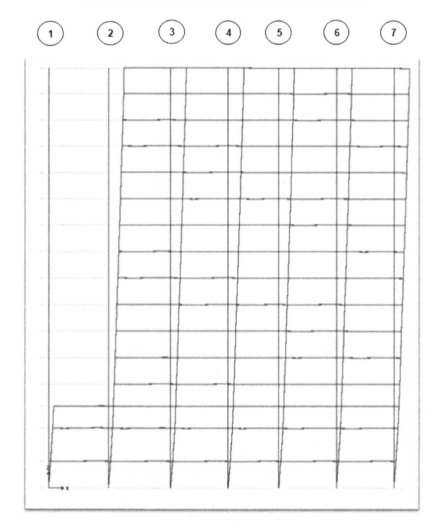

FIGURE 5.20   First modal shape.

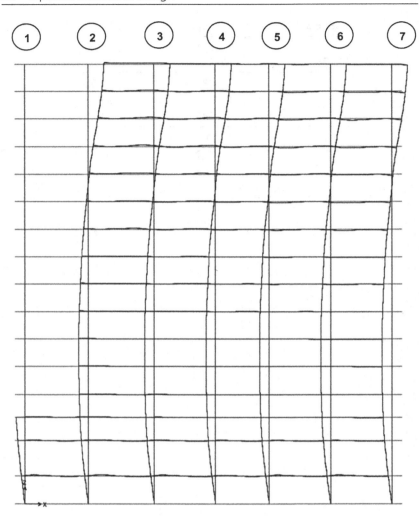

FIGURE 5.21   Second modal shape.

5.3.3 Example of Building with Bracings and a Hollow Space

The same building but with steel bracings is analyzed in a similar way as the other buildings. See an elevation view in Figure 5.29 and an isometric view in Figure 5.30.

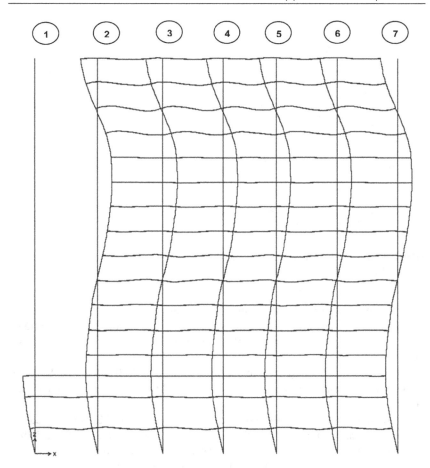

FIGURE 5.22 Third modal shape.

The building was designed under the current seismic code of Mexico City (GCDMX, 2014) and the cross-sections of columns, bracings, and beams are included in Tables 5.13, 5.14, and 5.15, respectively.

An elevation view with cross-section denominations for a typical frame appears in Figure 5.31.

The interaction diagrams for the columns were obtained and drawn, and samples for some of the critical columns are shown in Figures 5.32 and 5.33; in these Figures the grey point is the maximum demand point. Also, the corresponding maximum axial loads and moments, resistances, and ratios (P/Pr and M/Mr) appear in Table 5.16.

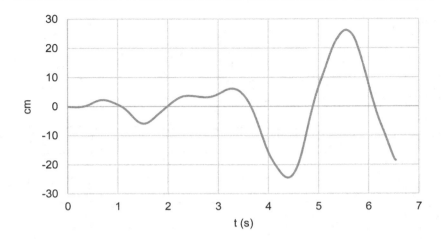

FIGURE 5.23   Roof lateral displacement for the first seconds of the 1985 Mexico City earthquake.

The maximum loads (moments, M, and shear forces, V) and resistances are calculated for beams; these values, locations, and ratios (V/Vr and M/Mr) for some of the critical beams, with cross-section 30cm × 40 cm, are shown in Table 5.17.

Bracings are located at levels 1 to 4 and bays A–B, B–C, D–E, F–G. Maximum loads and resistances for the critical bracings, and their locations, are shown in Table 5.18.

Interstory drifts under the east–west SCT record, 1985 Mexico City earthquake, were calculated to revise the serviceability limit state; see Table 5.19.

From Tables 5.16 to 5.19 it is observed that the dominant failure mode, once again, consists of the plastification of some beams, verifying once again the strong column-weak beam criteria. Shear mode is not significant.

The dynamic properties were also verified. Natural periods for the first three modes are shown in Table 5.20 and the first three modal shapes are in Figures 5.34, 5.35, and 5.36.

As expected, the first four floors, with bracings, almost do not move.

A distortion analysis is performed, for collapse prevention, and it was found that all of them satisfy the allowable interstory drift, 0.03 according to the current code in Mexico City (GCDMX, 2014). Table 5.21 shows the results.

As observed, bending on beams controls the failure, which indicates that the mechanism strong column-weak beam was followed in the design process. Therefore, the failure probability is assessed through the combination of local failures at the most critical beams, through Eq. (3.3) and, again, $PVF = 12.27$.

TABLE 5.9 Sample of Bending Moments (*tn-m*) and Stress Ratios for Critical Beams

LEVEL	AXIS 2	INTERSECTION	AXIS 3	INTERSECTION	AXIS 4	INTERSECTION	AXIS 5	INTERSECTION	AXIS 6
15	Mu-	64.53		61.28		30.22		61.27	
	Mr	299.53		299.53		299.53		299.53	
	Mu-/Mr	21.5%		20.5%		10.1%		20.5%	
	Mu+	35.07		27.76		8.69		28.64	
	Mr	176.46		176.46		176.46		176.46	
	Mu-/Mr	19.9%		15.7%		4.9%		16.2%	
6	Mu-	150.5		128.81		102.86		127.76	
	Mr	428.92		428.92		428.92		428.92	
	Mu-/Mr	35.1%		30.0%		24.0%		29.8%	
	Mu+	99.56		93.84		84.35		93.08	
	Mr	279.99		279.99		279.99		279.99	
	Mu-/Mr	35.6%		33.5%		30.1%		33.2%	
3	Mu-	156.4		164.79		147.92		157.7	
	Mr	547.78		547.78		547.78		547.78	
	Mu-/Mr	28.6%		30.1%		27.0%		28.8%	
	Mu+	111.58		129.02		125.61		122.21	
	Mr	434.48		434.48		434.48		434.48	
	Mu-/Mr	25.7%		29.7%		28.9%		28.1%	

Column C01 60 x 220 cm

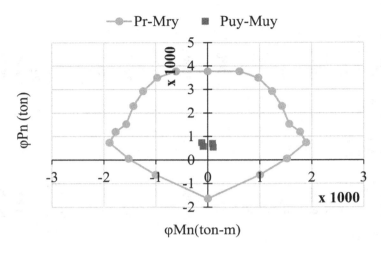

φMn(ton-m)

Condition	Pos.	Pu	Pr	Muy	Mr
		[Ton]	**[Ton]**	**[Ton*m]**	**[Ton*m]**
SX	Top	539.63	3767.97	103.94	1891.83
	Bottom	702.23	3767.97	92.33	1891.83
SX1	Top	570.59	3767.97	-85.19	-1892
	Bottom	734.12	3767.97	-113.27	-1891.83
		Pu/Pr	**Mu/Mr**	**D/C**	
SX	Top	0.143	0.055	**19.8%**	
	Bottom	0.186	0.049	**23.5%**	
SX1	Top	0.151	0.045	**19.6%**	
	Bottom	0.195	0.060	**25.5%**	

FIGURE 5.24 Interaction diagram and D/C ratios for column C01.

Four alternative designs were assessed and the optimal design was identified; Table 5.22 shows the results and the optimal design is in bold.

It is observed that the optimal reliability index is 3.81 and the optimal life-cycle cost is 13.28 million USD.

By following the same procedure as for the first building, some levels of epistemic uncertainty are introduced to assess the effect of these levels on the

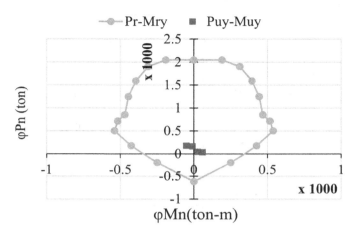

Column C03 60 x 130 cm

Puy-Muy

Condition	Pos.	Pu [Ton]	Pr [Ton]	Muy [Ton*m]	Mr [Ton*m]
SX	Top	35.97	503.61	59.36	538.67
	Bottom	164.98	2047.05	-9.64	-538.67
SX1	Top	44.51	714.34	26.49	517
	Bottom	178.09	1588.6	-50.57	-394.39

Condition	Pos.	Pu/pr	Mu/Mr	D/C
SX	Top	0.071	0.110	18.2%
	Bottom	0.081	0.018	9.8%
SX1	Top	0.062	0.051	11.4%
	Bottom	0.112	0.128	24.0%

FIGURE 5.25 Interaction diagram and D/C ratios for column C03.

building reliability index and expected life-cycle cost. The histograms for the reliability index and the expected life-cycle cost are shown in Figures 5.37 and 5.38.

The mean value, 90% and 95% percentiles are shown in Table 5.23.

Expected life-cycle costs for a 90% confidence β are shown in Figure 5.39.

TABLE 5.10 Interstory Drift Calculation for One Design of the Building

LEVEL	H cm	ΔH cm	MD cm	RD cm	ID
15	4930	330	21.7	0.4	0.001
14	4600	330	21.3	1.2	0.004
13	4270	330	20.1	1.4	0.004
12	3940	330	18.7	1.1	0.003
11	3610	330	17.6	0.9	0.003
10	3280	330	16.7	1.3	0.004
9	2950	330	15.4	1.1	0.003
8	2620	330	14.3	1.0	0.003
7	2290	330	13.3	1.4	0.004
6	1960	330	11.9	1.3	0.004
5	1630	330	10.6	1.0	0.003
4	1300	275	9.6	1.3	0.004
3	1025	275	8.3	1.1	0.004
2	750	415	7.2	1.2	0.005
1	335	335	5.9	3.0	0.007

H = height, ΔH = interstory height, MD = maximum displacement, RD = relative displacement, ID = interstory drift.

TABLE 5.11 Integration of Expected Life-Cycle Costs for 5 Design Alternatives

p_F	β	INITIAL	C_c	C_p	C_{fi}	C_F	$E(C_F)$	$E(LCC)$
0.0001	3.71	19.29	9.64	24	210	243.64	8.96E-06	19.29
0.0002	3.54	14.41	7.20	24	210	241.20	0.05919	14.47
2.19E-04	**3.51**	**12.15**	**6.07**	**24**	**210**	**240.07**	**0.64389**	**12.79**
3.59E-03	2.68	11.52	5.76	24	210	239.76	10.6	22.08
5.65E-03	2.53	11.08	5.54	24	210	239.54	16.6	27.69

5.3.4 Building with Walls and a Hollow Space

A similar case is studied with the same general dimensions of the building, except that some walls are added to improve its structural behavior. An isometric view is provided in Figure 5.40.

The plan views help to identify the general geometry of the building; see Figures 5.41 to 5.44 for some of them. The location of walls is indicated by the rectangle with dashed line.

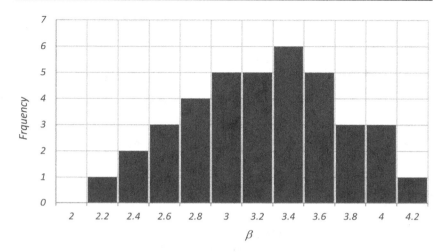

FIGURE 5.26   Histogram for reliability index.

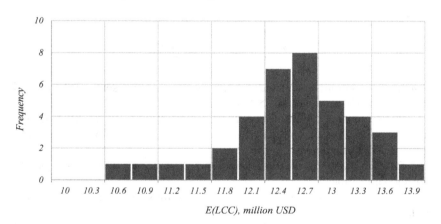

FIGURE 5.27   Histogram for expected life-cycle cost.

The elevation views help to identify the location of walls on the height; see Figures 5.45 and 5.46.

A distortion analysis is performed to revise the serviceability limit state; Table 5.24 shows the results. As observed, the code-allowable drift is satisfied and, therefore, the serviceability limit state is not critical.

A spectral response analysis is performed and the maximum load effects are calculated to identify the critical limit state. For the columns, the 2D interaction diagrams are obtained whereas, for the beams, the maximum acting and

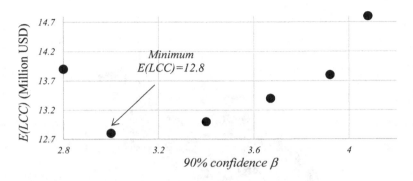

FIGURE 5.28 Expected life-cycle cost for 90% percentile of β.

TABLE 5.12 Mean Value
and 90 and 95 Percentiles
for Reliability Index

Mean	3.2
90%	3.8
95%	3.94

resistant moment are calculated. Interaction diagrams for some of the critical columns are shown in Figures 5.47 to 5.50.

Ratios between maximum forces and resistances for axial, bending, and shear forces were calculated to revise the limit state for columns (see Table 5.25) and beams.

Ratios between maximum moment and resistance for some critical beams are shown in Table 5.26.

As observed from the ratios, the bending limit state controls the building failure.

Similar to the other buildings, several alternative designs are prepared and the corresponding global reliabilities and costs are obtained. By following Eq. (5.2), the governing limit state for bending on the critical beams is assessed and the global failure probability and the expected life-cycle costs are calculated. A summary of the calculations is shown in Table 5.27.

In the same way as for the other building, the epistemic uncertainty is introduced and histograms of the reliability index and expected life-cycle costs are obtained. See Figures 5.51 and 5.52.

The mean value, 90% and 95% percentiles are shown in Table 5.28.

Expected life-cycle costs for a 90% confidence β are shown in Figure 5.53.

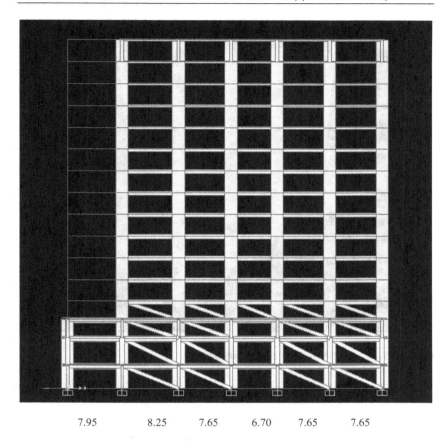

7.95   8.25   7.65   6.70   7.65   7.65

FIGURE 5.29   Elevation view (dimensions in m).

5.3.5 Steel Building with a Hollow Space

The same building, except that the design is made out of steel, is studied and a similar series of response analyses is performed to get the optimal design.

The columns shapes is a built-up "I" section made out of three plates; a sample of the shape and geometry is shown in Table 5.29. d is the depth, b_f the flange width, t_f the flange thickness, and t_w the web thickness. These dimensions were revised to avoid the slenderness that produces local buckling. In addition, all the sections were revised to prevent global buckling in columns and lateral buckling in beams.

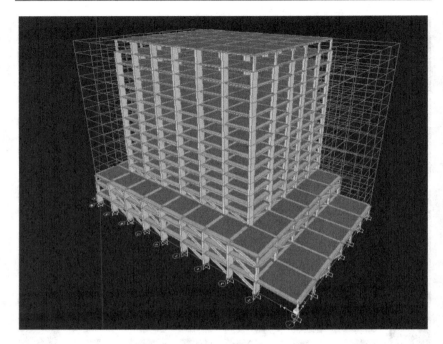

FIGURE 5.30   Isometric view of the building.

TABLE 5.13   Cross-Sections of Columns

DENOMINATION	DIMENSIONS (cm) b	h	LEVELS
C1	60	180	1, 2, and 3
C3	60	150	4, 5, and 6
C5	60	100	7, 8, and 9
C7	60	80	10, 11, and 12
C9	60	60	13 and 14
C10	45	45	15

TABLE 5.14   Bracings, Steel Box Sections

DENOMINATION	DIMENSIONS (cm) b	h	t
CV-1	20	20	1.905

TABLE 5.15 Beams Cross-Sections

| | DIMENSIONS (cm) | | |
DENOMINATION	b	h	LEVELS
T1	55	85	1, 2, and 3
T2	50	80	4, 5, and 6
T3	48	76	7 and 8
T4	46	72	9 and 10
T5	44	67	11 and 12
T6	42	62	13 and 14
T7	40	59	15

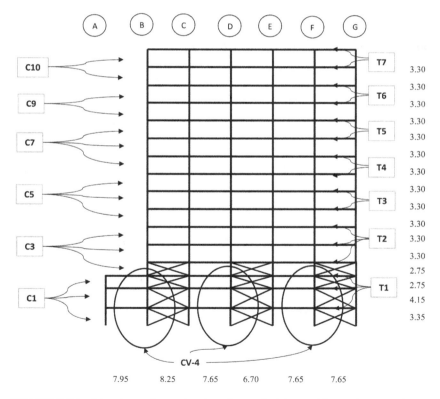

FIGURE 5.31 Members' denominations for typical frame (dimensions in m).

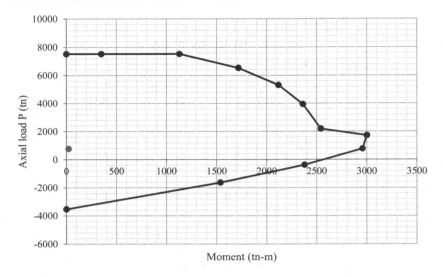

FIGURE 5.32   2D interaction diagram for column C1 60 × 180.

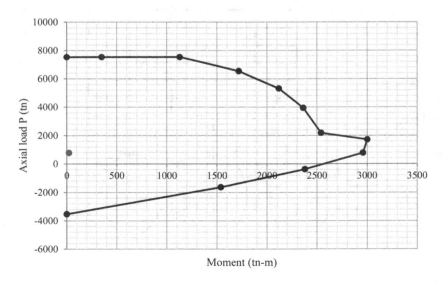

FIGURE 5.33   2D interaction diagram for column C3 60 × 150.

TABLE 5.16 Maximum Loads P (tn) and M (tn-m), Resistances Pr and Mr, Locations, and Ratios P/Pr and M/Mr for Critical Columns

COLUMNS (cm)	LOCATION (AXIS AND LEVEL L)	P	Pr	P/Pr	M	Mr	M/Mr
C1 110 × 180	C / L1, L2	774.888	7526.26	0.102	22.12	350	0.063
C3 110 × 165	C / L4, L5	577.674	7052.21	0.081	76.48	945	0.080
C5 90 × 140	C / L7, L8	368.522	4768.55	0.077	26.80	260	0.103
C7 75 × 75	C / L10, L11	180.383	2048.266	0.088	6.80	80.5	0.084

The cross-section for beams is a rectangular tube. See Table 5.30 for a sample of beam cross-sections. d, b and t are the depth, width, and thickness, respectively.

Plan and elevation views are shown in Figures 5.54 and 5.55. The elevation shows the frame in axis F and all the dimensions are in meters.

The same analyses are performed for this building as for the others, to identify the critical failure mode. Identically, given that the design seeks to satisfy the strong column-weak beam criterion, the critical limit state is the plastification of some beams.

Again, the calculation of the building failure probability follows Equation (5.2).

Additionally, the expected life-cycle costs are calculated for 4 alternative designs; a summary of these calculations is shown in Table 5.31.

Just as for the other buildings, the epistemic uncertainty is introduced and histograms of the reliability index and expected life-cycle costs are obtained. See Figures 5.56 and 5.57 for the histograms of the reliability index and the expected life-cycle cost for percentile 90, respectively.

The mean value, 90% and 95% percentiles are shown in Table 5.32.

Expected life-cycle costs for a 90% confidence β are shown in Figure 5.53.

TABLE 5.17 Maximum Loads V (tn) and M (tn-m), Resistances Vr and Mr, Locations, and Ratios V/Vr and M/Mr for some Beams

BEAM	LEVEL	# OF BAY	CROSS-SECTION	M	Mr	M/Mr	Vu	Vr	V/Vr
1	1	1	30 × 40	12.5	60.7	0.20	14.3	116.5	0.12
2	1	2	30 × 40	14.31	60.7	0.23	15.5	116.5	0.13
7	2	1	30 × 40	12.83	60.7	0.21	14.4	116.5	0.12
8	2	2	30 × 40	14.44	60.7	0.23	15.5	116.6	0.13
13	3	1	30 × 40	14.57	60.6	0.24	16.3	116.0	0.14
14	3	2	30 × 40	14.64	60.7	0.24	15.6	116.6	0.13
44	9	1	30 × 50	19.8	78.9	0.25	17.5	123.3	0.14
45	9	2	30 × 50	16.0	78.9	0.20	15.2	123.6	0.12
59	12	1	30 × 50	21.6	78.9	0.27	18.0	123.1	0.14
60	12	2	30 × 50	17.9	78.9	0.22	15.8	123.5	0.12

TABLE 5.18 Maximum Axial Loads P (tn) and Resistances Pr (tn), Locations, and Ratios P/Pr for Critical Bracings

BRACINGS	LOCATION (LEVEL L AND AXIS)	P	Pr	P/Pr
C-V 4	L2–L3/A–B	25.64	401.34	0.063

TABLE 5.19 Interstory Drift for Record SCT-EW During 1985 Mexico City Earthquake

LEVEL	DISPLACEMENT (cm)	HEIGHT (cm)	RELATIVE DISPLACEMENTS (cm)	INTERSTORY DRIFT
15	5.5246	330	0.3864	0.0011709
14	5.1382	330	0.4017	0.0012173
13	4.7365	330	0.4923	0.0014918
12	4.2442	330	0.4933	0.0014948
11	3.7509	330	0.5291	0.0016033
10	3.2218	330	0.5552	0.0016824
9	2.6666	330	0.5247	0.00159
8	2.1419	330	0.5231	0.0015852
7	1.6188	330	0.4993	0.001513
6	1.1195	330	0.3981	0.0012064
5	0.7214	330	0.3144	0.0009527
4	0.4070	275	0.1328	0.0004829
3	0.2742	275	0.0956	0.0003476
2	0.1786	415	0.1223	0.0002947
1	0.0563	335	0.0563	0.0001681

TABLE 5.20 First 3 Natural Periods

MODE	PERIOD (sec)
1	0.966
2	0.323
3	0.176

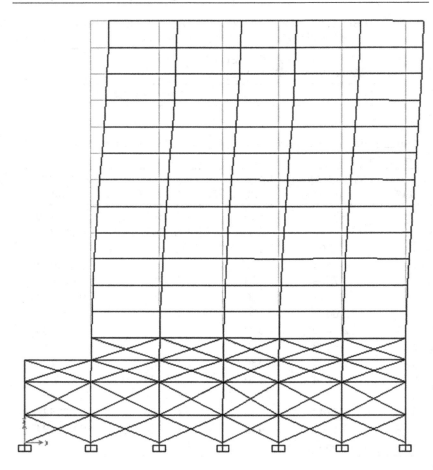

FIGURE 5.34   First modal shape.

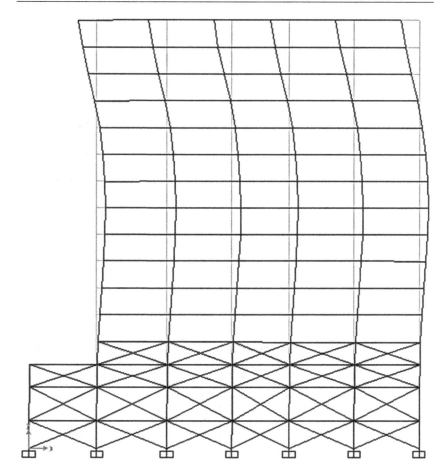

FIGURE 5.35 Second modal shape.

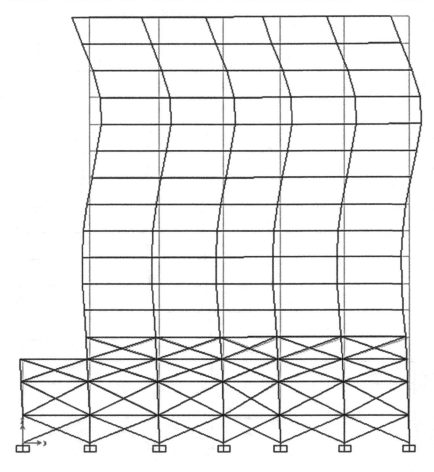

FIGURE 5.36 Third modal shape.

TABLE 5.21   Interstory Drift Calculation for One Design of the Building

LEVEL	H cm	MD cm	ΔH cm	RD cm	ID
15	4930	23.3	330	1.4	0.0043
14	4600	21.9	330	1.9	0.0059
13	4270	19.9	330	2.5	0.0075
12	3940	17.5	330	2.6	0.0080
11	3610	14.8	330	2.7	0.0081
10	3280	12.2	330	2.6	0.0078
9	2950	9.6	330	2.4	0.0071
8	2620	7.2	330	2.1	0.0063
7	2290	5.2	330	1.8	0.0055
6	1960	3.3	330	1.5	0.0045
5	1630	1.8	330	1.0	0.0030
4	1300	0.9	275	0.4	0.0013
3	1025	0.5	275	0.2	0.0008
2	750	0.3	415	0.2	0.0005
1	335	0.1	335	0.1	0.0002

H = height, ΔH = interstory height, MD = maximum displacement, RD = relative displacement, ID = interstory drift.

TABLE 5.22   Integration of Expected Life-Cycle Costs for 4 Design Alternatives

p_F	β	INITIAL	C_c	C_p	C_{fi}	CF	E(CF)	E(LCC)
0.000009	4.29	21.55	10.78	24.00	210.00	244.78	0.02	21.58
0.000070	**3.81**	**13.10**	**6.55**	**24.00**	**210.00**	**240.55**	**0.18**	**13.28**
0.000594	3.24	12.59	6.30	24.00	210.00	240.30	1.75	14.34
0.002347	2.83	12.15	6.08	24.00	210.00	240.08	6.91	19.06

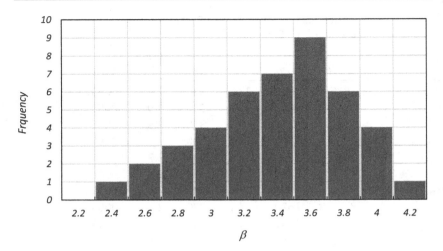

FIGURE 5.37   Histogram for reliability index.

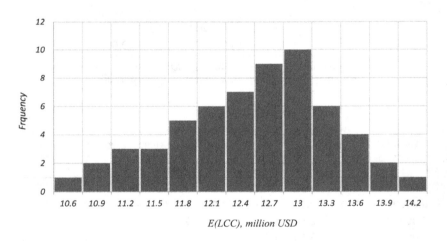

FIGURE 5.38   Histogram for expected life-cycle cost.

TABLE 5.23   Mean Value and 90 and 95 Percentiles for Reliability Index

Mean	3.36
90%	3.84
95%	3.96

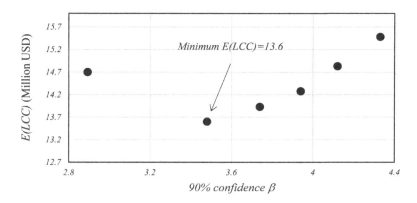

FIGURE 5.39   Expected life-cycle cost for 90% percentile of β.

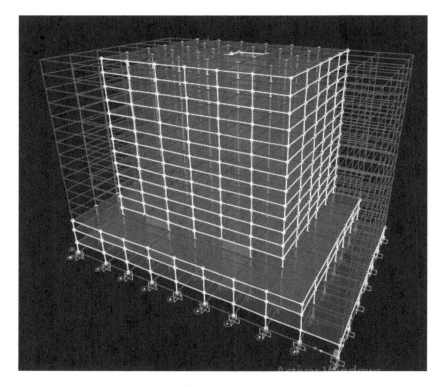

FIGURE 5.40   Isometric view of building with walls.

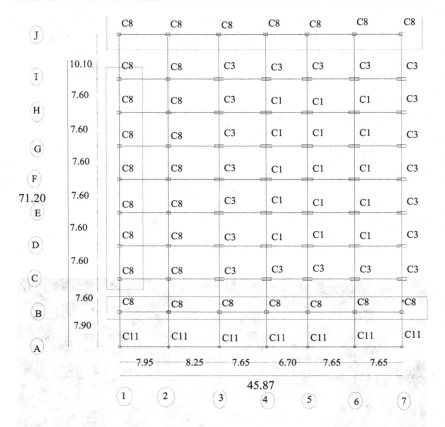

FIGURE 5.41   Plan view level 1 (dimensions in m).

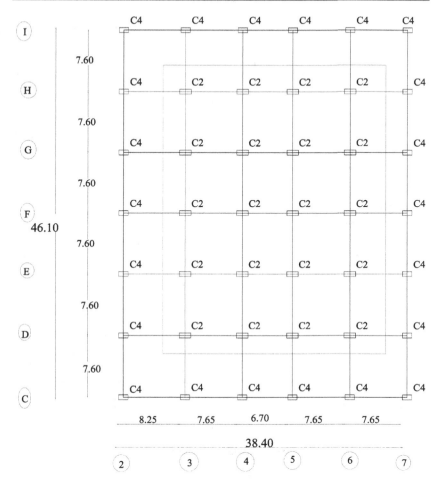

FIGURE 5.42 Plan view levels 5 to 8 (dimensions in m).

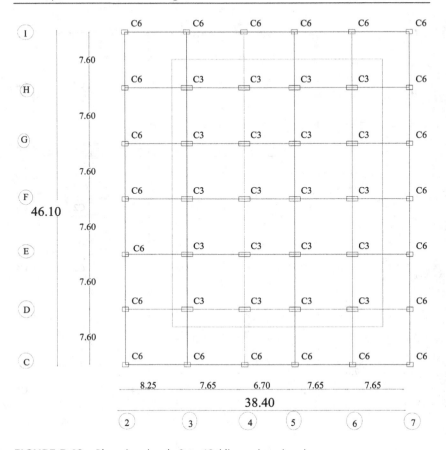

FIGURE 5.43   Plan view levels 9 to 12 (dimensions in m).

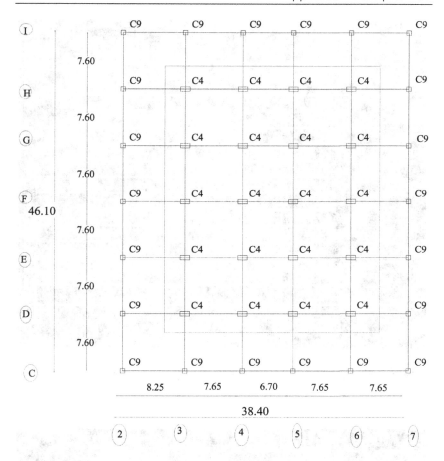

FIGURE 5.44 Plan view levels 13 to 15 (dimensions in m).

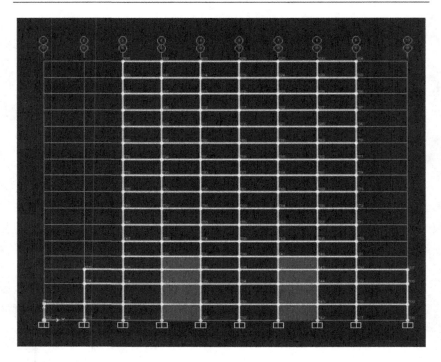

FIGURE 5.45   Elevation view of frame 4.

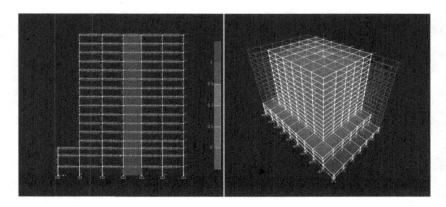

FIGURE 5.46   Elevation view of frame *F* and isometric view.

TABLE 5.24 Distortion Analysis, Dimensions in cm

LEVEL	DISPLACEMENT	HEIGHT	RELATIVE DISPLACEMENT	DISTORTION
15	7.43	330	0.34	0.0011
14	7.08	330	0.42	0.0013
13	6.65	330	0.50	0.0015
12	6.14	330	0.54	0.0017
11	5.59	330	0.58	0.0018
10	5.01	330	0.62	0.0019
9	4.39	330	0.64	0.0019
8	3.74	330	0.63	0.0019
7	3.11	330	0.61	0.0019
6	2.49	330	0.59	0.0018
5	1.90	330	0.51	0.0015
4	1.39	275	0.31	0.0011
3	1.08	275	0.25	0.0009
2	0.83	415	0.43	0.0011
1	0.39	335	0.39	0.0012

TABLE 5.25 Ratios Between Maximum Force and Resistance, for Axial Force, Moments and Shear Forces, for Critical Columns

COLUMN	LEVEL	BAY	Mmax tn-m	Mr tn-m	Mmax/ Mr	Pmax tn	Pr tn	Pmax/ Pr	MAXIMUM SHEAR tn	RESISTANT SHEAR tn	Vmax/ Vr
1	4	3	82.7	1,266.3	0.04	1018.5	4767.1	0.21	30.1	195.3	0.15
2	8	3	49.7	468.3	0.06	660.9	2889.1	0.23	26.0	119.2	0.22
3	4	3	62.3	1,386.1	0.03	828.9	4731.0	0.18	25.4	184.5	0.14
4	6	3	27.0	680.8	0.03	449.8	2618.4	0.17	10.6	103.2	0.10
8	1	2	14.6	240	0.25	93.6	1950	0.05	10.4	76.2	0.13

TABLE 5.26 Ratios Between Maximum Moment and Resistance, for some Critical Beams

MEM-BER NO.	LEVEL	BAY	DIMENSIONS (m) b	DIMENSIONS (m) h	BENDING RESISTANCE tn-m	MAXIMUM MOMENT tn-m	BENDING RATIO	SHEAR RESISTANCE tn	MAXIMUM SHEAR tn	SHEAR RATIO
7	1	3	0.4	0.57	15.6	75.6	0.206	11.34	65.25	0.173
16	2	3	0.4	0.57	15.3	75.6	0.203	11.37	65.25	0.174
23	3	3	0.4	0.57	15.6	75.6	0.206	11.34	65.25	0.173
30	4	3	0.4	0.57	13.4	75.6	0.177	9.66	65.25	0.148
36	5	3	0.4	0.57	14.8	75.6	0.196	11.21	65.25	0.171
42	6	3	0.4	0.56	14.8	75.6	0.196	11.17	65.25	0.171
52	7	3	0.4	0.56	14.8	75.6	0.195	11.17	65.25	0.171
58	8	3	0.4	0.56	15.5	75.6	0.204	11.41	65.25	0.174

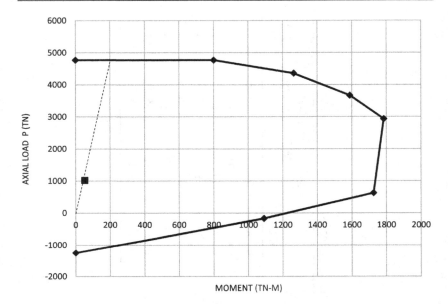

FIGURE 5.47   Interaction diagram for column C1 (120cm × 120cm). The black square is the acting load.

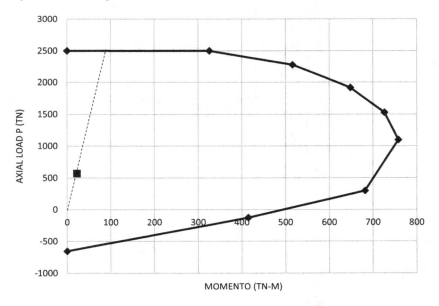

FIGURE 5.48   Interaction diagram for column C3 (80cm × 173cm). The black square is the acting load.

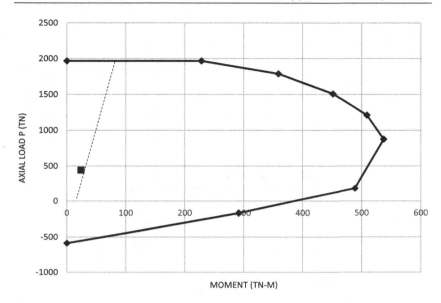

FIGURE 5.49 Interaction diagram for column C4 (80cm × 155cm). The black square is the acting load.

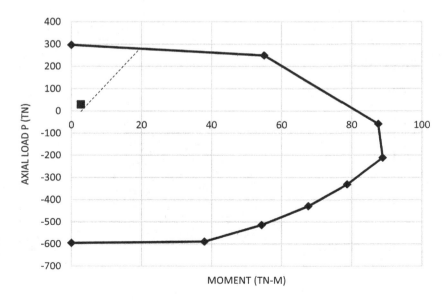

FIGURE 5.50 Interaction diagram for column C8 (60cm × 65cm). The black square is the acting load.

TABLE 5.27 Integration of expected life-cycle costs (million USD) for 4 alternative designs

p_F	β	INITIAL	C_c	C_p	C_{fi}	CF	E(CF)	E(LCC)
1.4E-05	4.18	19.6	9.8	24	210	243.8	4.18E-02	19.64
1.5E-04	3.61	19	9.5	24	210	243.5	4.49E-01	19.44
5.3E-04	3.27	18.8	9.4	24	210	243.4	1.60E+00	20.47
5.2E-03	2.56	21.8	10.9	24	210	244.9	1.57E+01	37.52

TABLE 5.28 Mean Value and 90 and 95 Percentiles for Reliability Index

Mean	3.34
90%	3.7
95%	3.9

TABLE 5.29 Sample of Designation and Geometries for 3 Columns

TAG	SECTION SHAPE	d IN (mm)	t_w IN (mm)	b_f IN (mm)	t_f IN (mm)
C-1	3 plates "I"	36 (900)	1 1/8 (28.6)	24 (600)	1 ½ (38.1)
C-2	3 plates "I"	36 (900)	1 (25.4)	24 (600)	1 ¼ (31.8)
C-3	3 plates "I"	36 (900)	1 (25.4)	20 (500)	1 ¼ (31.8)

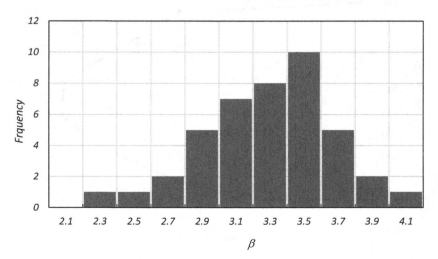

FIGURE 5.51 Histogram of reliability index for building with walls.

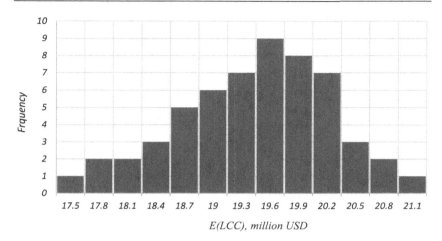

FIGURE 5.52 Histogram of expected life-cycle costs.

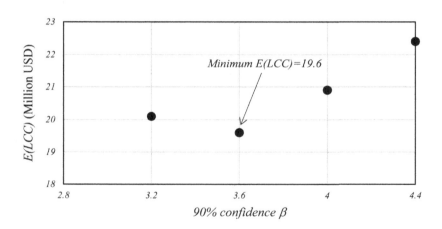

FIGURE 5.53 Expected life-cycle cost for 90% percentile of β.

TABLE 5.30 Sample of Three Cross-Sections for Beams

TAG	SECTION SHAPE	d (mm)	t (mm)	b (mm)
T-1	HSS 16 × 8 × 3/8"	406	8.9	203
T-2	HSS 16 × 8 × 1/2"	406	11.8	203
T-3	HSS 16 × 12 × 3/4"	406	18.9	304

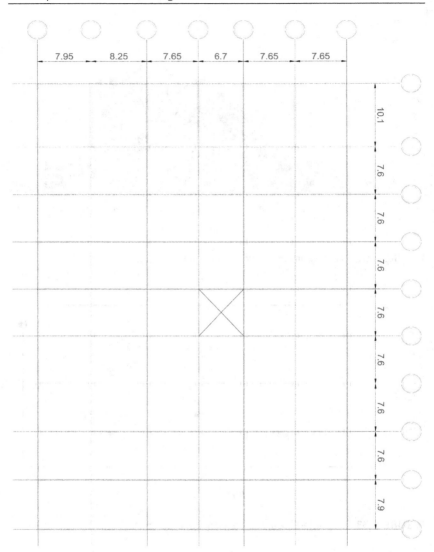

FIGURE 5.54 Plan view, dimensions in m.

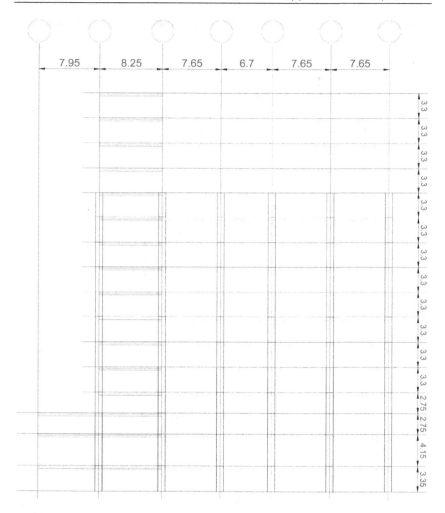

FIGURE 5.55 Elevation view of frame *F*, dimensions in m.

TABLE 5.31   Integration of Expected Life-Cycle Costs (million USD) for 4 Alternative Designs

p_F	β	INITIAL	C_C	C_P	C_{FI}	CF	E(CF)	E(LCC)
1.4E-05	4.18	19.6	9.8	24	210	243.8	4.18E-02	19.64
1.5E-04	3.61	19	9.5	24	210	243.5	4.49E-01	19.44
5.3E-04	3.27	18.8	9.4	24	210	243.4	1.60E+00	20.47
5.2E-03	2.56	21.8	10.9	24	210	244.9	1.57E+01	37.52

TABLE 5.32   Mean value and 90 and 95 percentiles for reliability index

Mean	3.34
90%	3.7
95%	3.9

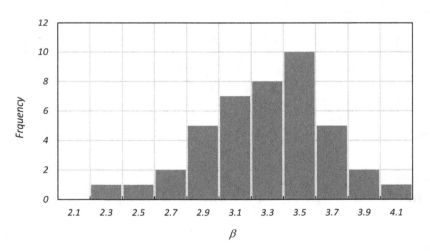

FIGURE 5.56   Histogram of reliability index for building with walls.

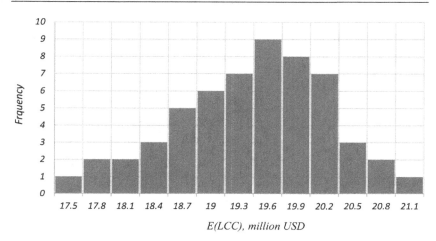

FIGURE 5.57 Histogram of expected life-cycle costs.

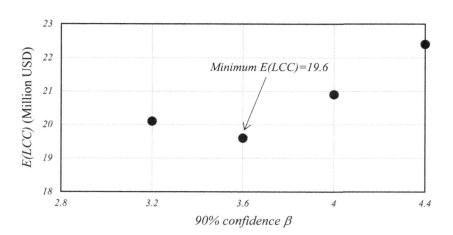

FIGURE 5.58 Expected life-cycle cost for 90% percentile of β.

Analysis of Results

6

The incorporation of epistemic uncertainty into the safety analysis of infrastructure and large structures, such as medium- and high-rise buildings, provides an excellent opportunity to extend the field for decision making, in particular for optimal design, to offer designers and investors several degrees of conservativeness through percentiles in addition to the mean values.

For large infrastructure systems, like bridges or marine platforms, the loss of benefits derived from the interruption of services largely offset the cost of the structure itself and, therefore, there should be no restrictions to investment in initial, repair, or maintenance costs to keep the system in adequate operating condition.

In the case of buildings, the cost of potential fatalities is the largest cost component, because of the high density of occupation. The results showed that the optimal reliability index is between 3.3 and 3.8, which corresponds to a 110–130% overdesign in respect to the code-recommended design for Mexico City. It is observed that, when the building has a hollow space, a reinforced design of the neighboring frames is required compared to the requirements for a case with no hollow space; also, the costs increase. The building with bracings increased its cost because the bracings concentrate more axial load on the foundation, and that means stronger and more expensive supports. The building with walls is also more expensive than the building with only frames, although less expensive than the one with bracings. The criteria strong column-weak beam means that the beams tend to fail before the columns, and this is more convenient from the point of view of overall safety. The limit state consisted of the combination of a series of local failures on the beams, i.e. those that showed the highest ratio between maximum moment and bending resistance. Brittle and buckling failure modes were prevented for the concrete buildings and the steel building. For optimal design, these principles should be maintained and the incorporation of epistemic uncertainty should be promoted; otherwise, large and complex infrastructure systems and buildings with irregularity of geometry and/or stiffness may produce torsional problems.

The use of 90% confidence, or percentile, for the reliability index and the expected life-cycle cost is close to the professional practice for infrastructure

DOI: 10.1201/9781003177289-6

systems and for buildings under regular geometry. For buildings with hollow spaces, torsional problems in seismic environments move the optimal reliability to slightly higher values, by 10 or 15%.

The probability density evolution method (PDEM) showed its effectiveness in all cases, particularly when the structure possesses special features, like irregularities that require more simulations; they may be treated more easily with PDEM than with crude MC simulation. For cases that require detailed calculations, such as when conservative decisions are under examination for important infrastructures and large or complex buildings, the advantages of PDEM become more evident.

Conclusions and Recommendations 7

The main conclusions of this study may be summarized as follows:

- The required safety in the design of a structural system cannot be prescribed in the way of a structural component. Each structural system is unique; thus, the required safety for its design must be determined independently for each system.
- A procedure for determining the safety index for the optimal design of a structural system is proposed; the procedure is based on achieving the minimum expected life-cycle cost design of the structure.
- The probability density evolution method (PDEM) is an effective method for the response assessment of complex systems. The method is based on mathematical rigor, and represents an effective alternative to the Monte Carle simulation (MCS) for complex systems.
- As illustrated in the example of a 15-story reinforced concrete (RC) building, the proposed procedure shows that the building in Mexico City could have been designed for earthquake resistance with higher safety and at lower cost (with a saving of at least around 0.5 million USD).
- The optimal reliability index for the 15-story building with no hollow spaces is about 3.4, whereas for buildings with a hollow space it is between 3.5 and 3.8 depending on the use of frames, bracings, or walls as a structural system.
- For infrastructure systems, the optimal design strongly depends on the value of the provided service or the revenues produced by the operation whereas, for buildings, depends on the potential fatalities or the occupation density.
- For the building with a hollow space, the use of bracings produced a higher reliability and a higher cost when compared to the framed building. But the use of a limited number of walls is less expensive than the use of bracings. This is because the bracings require more structural additions on the foundation than those required for walls.

DOI: 10.1201/9781003177289-7

- It is recommended to pursue the development and applications of the PDEM because of its advantages for performing an efficient simulation process, especially in situations where detailed calculations, multiple hazards, or conservative decisions are under examination.

- It is recommended to explore the potential of PDEM to identify the benefits, from the perspective of the information value, of health monitoring and lifetime extension for instrumented infrastructure or important buildings to develop optimal risk-based inspection/ maintenance schedules.

References

American Petroleum Institute (API) (1993). Recommended practice for planning, designing and constructing fixed offshore platforms: Load and resistance factor design. In: *API RP 2ª LRFD*, 1st Ed., July, API, Washington, DC.

Ang A.H.-S. & Ma H.-F. (1981). On the reliability of structural systems. In: Proceedings of the 3rd ICOSSAR, Trondheim, Norway.

Ang A.H.-S., De Leon D. & Fan W. (2019). Optimal reliability-based aseismic design of high-rise buildings. *Structure and Infrastructure Engineering* 16(4):520–530. DOI: 10.1080/15732479.2019.1653327

Bai Y. & Jin W.-L. (2016). *Marine Structural Design*, 2nd Ed. ISBN-13: 978-00809 99975, ISBN-10: 9780080999975.

Chakri A., Yang X.-S., Khelif R. & Becouaret M. (2017). Reliability based design optimization using the directional bat algorithm. *Neural Computing and Applications*, 30: 2381–2402. First published online, 2017. DOI: 10.1007/s00521-016-2797-3

Chen J.B. & Li J. (2007). Development-process-of-nonlinearity-based reliability evaluation of structures. *Probabilistic Engineering Mechanics* 22:267–275.

Chen J.B. & Wan Z.Q. (2019). A compatible probabilistic framework for quantification of simultaneous aleatory and epistemic uncertainty of basic parameters of structures by synthesizing the change of measure and change of random variables. *Structural Safety* 78:76–87.

Chen J.B., Ghanem R. & Li J. (2009). Partition of the probability-assigned space in probability density evolution analysis of nonlinear stochastic structures. *Probabilistic Engineering Mechanics* 24:27–42.

Chen J., Yang J. & Jensen H. (2020). Structural optimization considering dynamic reliability constraints via probability density evolution method and change of probability measure. *Structural and Multidisciplinary Optimization*, 62: 2499–2516. DOI: 10.1007/s00158-020-02621-4

De Leon D. & Ang A.H-S. (2008). Confidence bounds on structural reliability estimations for offshore platforms. *Journal of Marine Science and Technology* 13(3): 308–315.

Der Kiureghian A. (2008). Analysis of structural reliability under parameter uncertainties. *Probabilistic Engineering Mechanics* 23:351–358. DOI: 10.1016/j.probengmech.2007.10.011

Dong Y. & Frangopol D. (2015). Risk-informed life-cycle optimum inspection and maintenance of ship structures considering corrosion and fatigue. *Ocean Engineering* 101: 161–171. DOI: 10.1016/j.oceaneng.2015.04.020

Dueñas-Osorio L. & Mohan V.S. (2009). Cascading failures in complex infrastructure systems. *Structural Safety* 31(2):157–167. DOI: 10.1016/j.strusafe.2008.06.007

Ellingwood B. & Galambos T.V. (1982). Probability-based criteria for structural design. *Structural Safety* 1(1): 15–26.

Ellingwood B.R. & Dusenberry D.O. (2005). Building design for abnormal loads and progressive collapse. *Computer-Aided Civil and Infrastructure Engineering* 20(3):194–205. DOI: 10.1111/j.1467-8667.2005.00387.x

Ellingwood B.R., Smilowitz R., Dusenberry D.O. et al. (2007). *Best Practices for Reducing the Potential for Progressive Collapse in Buildings*, National Institute of Standards and Technology, US Department of Commerce, Washington, DC, NIST/R7346.

Fan W., Chen J.B. & Li J. (2009). Solution of generalized evolution equation via a family of δ sequences. *Computational Mechanics* 43:781–796.

Francis R. & Bekera B. (2014). A metric and frameworks for resilience analysis of engineered and infrastructure systems. *Reliability Engineering and System Safety* 121(C):90–103. DOI: 10.1016/j.ress.2013.07.004

Frangopol D.M. & Liu M. (2007). Maintenance and management of civil infrastructure based on condition, safety, optimization, and life-cycle cost. *Structure and Infrastructure Engineering* 3(1):29–41. DOI: 10.1080/15732470500253164

Frangopol D.M., Kawatani M. & Kim Ch.W. (2007). *Reliability and Optimization of Structural Systems: Assessment, Design, and Life-Cycle Performance*, CRC Press, Boca Raton, FL. 280 Pages, ISBN 9780415406550

Ghosn M., Dueñas-Osorio L., Frangopol D. & Mcallister T. (2016). Performance indicators for structural systems and infrastructure networks. *Journal of Structural Engineering* 142(9):F4016003. DOI: 10.1061/(ASCE)ST.1943-541X.0001542

Gobierno de la Ciudad de México (GCDMX) (2014). *Normas Técnicas Complementarias para Diseño por Sismo en la Ciudad de México* (In Spanish), GCDMX, México.

Han S.H. & Ang A.H.-S. (2008). Optimal design of cable-stayed bridges based on minimum life-cycle cost. In: Proceedings of the IABMAS'08, Seoul, Korea.

Hyeon Ju B. & Chai Lee B. (2008). Reliability-based design optimization using a moment method and a kriging metamodel. *Journal Engineering Optimization* 40(5):421–438. DOI: 10.1080/03052150701743795

Lange D., Honfi D., Theoharidou M., Giannopoulos G., Kristina N. & Storesund K. (2017). Incorporation of resilience assessment in critical infrastructure risk assessment frameworks. In: 2nd International Conference on Engineering Sciences and Technologies, Slovenia. DOI: 10.1201/9781315210469-132

Li J. (2018). A PDEM-based perspective to engineering reliability: From structural elements to lifeline network. In: Proceedings of the 6th International Symposium on Reliability Engineering and Risk Management (6ISRERM), Singapore.

Li J. & Chen J.B. (2008). The principle of preservation of probability and the generalized density evolution equation. *Structural Safety* 30:65–77.

Li J. & Chen J.B. (2009). *Stochastic Dynamics of Structures*, 103(GT11):1227–1246.

Li J., Chen J.B., Sun W.L. & Peng Y.B. (2012). Advances of the probability density evolution method for nonlinear stochastic systems. *Probabilistic Engineering Mechanics* 28:132–142.

Petróleos Mexicanos (PEMEX) (2000). *Diseño y Evaluación de Plataformas Marinas Fijas en la Sonda de Campeche*. NRF-003-PEMEX-2000, Rev. 0, Committee of Normalization for PEMEX and Subsidiaries Organisms.

Rajan A., Luo F., KuangY.Ch. & Ooi M. (2020). Reliability-based design optimisation of structural systems using high-order analytical moments. *Structural Safety* 86:101970. DOI: 10.1016/j.strusafe.2020.101970

Renschler Ch.S., Frazier A.E., Lucy Arendt L. & Cimellaro G.P. (2010). Developing the 'PEOPLES' resilience framework for defining and measuring disaster resilience at the community scale. In: 9th US and 10th Canadian Conference on Earthquake Engineering, Toronto, Canada. DOI: 10.13140/RG.2.1.1563.4323

Santander C.F. & Sanchez-Silva M. (2008). Design and maintenance programme optimization for large infrastructure systems. *Structure and Infrastructure Engineering* 4(4):297–309. DOI: 10.1080/15732470600819104

Stochino F., Bedon Ch., Sagaseta J. & Honfi D. (2019). Robustness and resilience of structures under extreme loads. *Advances in Civil Engineering* 2019(23–24):1–14. DOI: 10.1155/2019/4291703

Thoft-Christensen P. (1991). On reliability-based structural optimization. In: Der Kiureghian A., Thoft-Christensen P. (eds) *Reliability and Optimization of Structural Systems '90.* Lecture Notes in Engineering, vol 61, Springer, Berlin, DOI: https://doi.org/10.1007

Xu H. & Gardoni P. (2020). Multi-level, multi-variate, non-stationary, random field modeling and fragility analysis of engineering systems. *Structural Safety* 87:101999. https://doi.org/10.1016/j.strusafe.2020.101999

Zhang Z. & Lu D.-G. (2018). Global reliability analysis of RC frame structures under earthquakes using subset simulation. Proc. of the 6th Intl. Symposium on Reliability Engineering and Risk Management (6ISRERM), Singapore.

Index

Printed in the United States
by Baker & Taylor Publisher Services